THE FUTURE OF Communication

By Diane Bailey

CREATIVE EDUCATION

TABLE OF Contents

INTRODUCTION

Snow days may soon be a thing of the past, as students in the future might not have to actually go to school. Instead, they'll be able to send a hologram. Through a few tricks of light, they can be anywhere at any time—or at least look as though they are. If they want to be there in person but are running late, that won't be a problem, either. Global Positioning System (GPS) devices in their shoes could automatically call their friends' cell phones. If students need to research a project, they can tell their computer, which will handle much of the work automatically. With compact phones that will fit in their pockets yet contain the power of several of today's computers, they'll be able to communicate with their family and friends—and even their homes—at any time.

Such a scenario has not happened yet, but the research to make it a reality is underway. Scientists are working to make computers faster and smaller and investigating how to transfer human intelligence to machines. They are developing tiny materials that could mean big changes in how communications devices are built and operated. Meanwhile, companies and governments are figuring out how to spread this technology around the world. What kinds of communication will people want? New technologies sometimes create as many problems as they solve. Always being available could be convenient, but it could also be **intrusive**. In the future, staying in touch won't be the problem—staying out of it will be!

Cell phone communication is likely to soon offer the option of holographic images (above), as computer parts continue to shrink in size yet grow in power

CHAPTER ONE 1 CHAPTER ONE

MAKING A CONNECTION

During the American Civil War (1861–65), United States president Abraham Lincoln did not wait for important telephone calls to update him on battles, as the telephone had not been invented yet. However, he did set up a cot in an office with a telegraph—the most advanced form of communications technology at the time. Lincoln's secretaries wrote that "his thoughts by day and anxiety by night fed upon the intelligence [information] which the telegraph brought."

Whether it is working, playing, or waging war, the things people do depend on how—and how well—we can communicate. For centuries, the primary way for humans to communicate was through spoken words. Then, around 6500 B.C., people began using written symbols. During a period called the Bronze Age, the cultures of the ancient Sumerians and Egyptians developed complex alphabets and writing systems. Writing represented an advance because the receiver did not have to be present at the exact moment the message was being created. Also, it could be spread to multiple people. Today, people like "real-time" communication. But thousands of years ago, they just wanted a pen.

The first books and documents were written by hand. Only a few people knew how to write, and the process was long and laborious. Then, in the 1400s, a German printer named Johannes Gutenberg invented a printing press that used movable type. Instead of having to write each document separately, a person could simply move the type into place, smear it with ink, and press

The printing press of Johannes Gutenberg (above) ushered in a new era of mass communication, and the telegraph (top) revolutionized long-distance communication

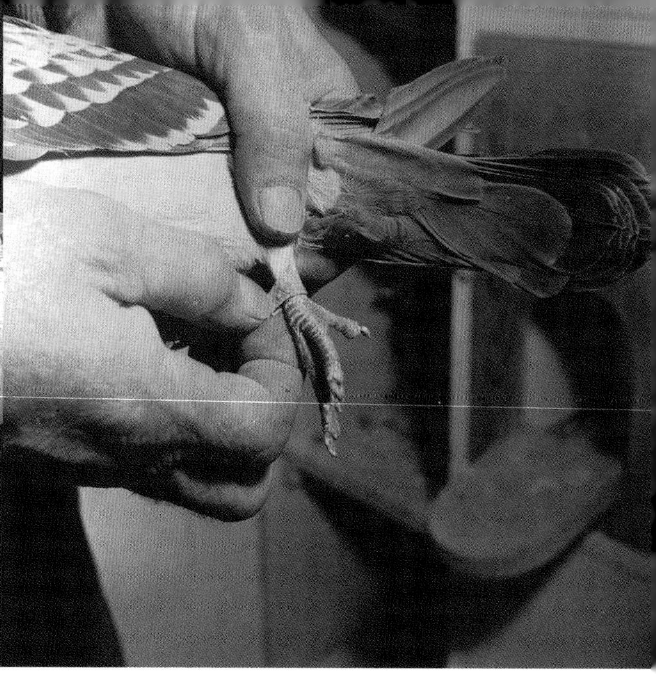

it onto paper. Soon, books were being printed and distributed throughout **medieval** Europe. Many improvements have been made to the printing press in the last 500 years, but the basic concept is the same, and printing presses are still used today. Their invention started "mass communication" as we now know it.

Newspapers were being published at least as early as 59 B.C., in Rome, Italy, but they did not become widespread until almost 1,500 years later, when better printing methods came along. The number of newspapers and frequency of their publication grew quickly. In 1830, about 700 newspapers were being published in the U.S. Forty years later, that number had grown to more than 5,000. Newspapers, magazines, and, later, radio and television brought information to large numbers of people. Government, the economy, scientific exploration—all were transformed by the spread of knowledge to the public.

People sometimes laugh about the speed of "snail mail" today, but before telephones and e-mail, mail was a lifeline to the world. Postal services of various forms took the burden off an individual to deliver his own message. Carrier pigeons were used thousands of years ago to deliver messages that were placed in tubes tied around the birds' necks or legs. Pigeons proved to be great timesavers. A person had to take the time to write a message, but he was off the hook in terms of having to actually deliver it. Even after more advanced technology had arrived, in the mid-1800s, German media **mogul** Paul Reuter was using pigeons to send news and stock prices to telegraph

From left to right, images of past communication: newsboys selling papers, a pigeon being prepared to carry a message, and workers in a telegraph exchange office

offices, from where they were more widely distributed. Ancient civilizations in China and Rome used systems of relay stations, transferring a message from one human carrier to another along an established route so no one person had to travel too far. In mid-19th-century America, the famous Pony Express was the very image of fast, reliable mail service. However, even a fast horse was snaillike compared with the up-and-coming technology called the telegraph.

By the mid-1700s, scientists were starting to better understand how electricity worked. Several scientists and inventors came up with devices that would transmit electrical signals over wires. In 1832, American inventor Samuel Morse, along with his assistant Alfred Vail, refined the technology so that signals could be sent over several miles rather than just a few hundred yards. Perhaps the duo's greatest contribution, however, was Morse code, a system of dots and dashes that represented letters and could be easily transmitted over the wires to telegraphs stationed in different cities.

Even 150 years ago, however, the limitations of wires in communication were obvious. The main drawback was that a severed wire meant a halt in communication. In the late 1800s, Italian inventor Guglielmo Marconi demonstrated that signals could be sent using the electrical power in radio waves, with no need for wires at all. In 1901, Marconi sent a radio transmission across the Atlantic Ocean. Ten years later, when the famous ocean liner *Titanic* was sinking in the Atlantic, it was wireless telegraphy that allowed the ship to send a distress call.

The photo opposite captures Alexander Graham Bell in 1892 as he placed the inaugural call opening a telephone connection between New York and Chicago

Meanwhile, breakthroughs were being made with wires, too. In 1876, Scottish-born inventor Alexander Graham Bell demonstrated his new device, the telephone. The telegraph had basically provided a way of sending long-distance messages using an audio code. But the telephone was faster and more user-friendly, as people could skip the "code" part and just talk.

At first, it was not clear whether the telephone would catch on. The sound quality was poor, and there were no switchboards through which to route calls. The telephone was seen as an interesting gadget but not an essential one. The market slowly grew, however, and soon, transatlantic cables were laid under the ocean to make international calls possible. In 1962, through an international **venture** among government agencies in the U.S., Great Britain, and France, a communications **satellite** called TELSTAR was launched into orbit around Earth. Satellites act sort of like mirrors, as a signal is projected to the satellite and then "bounced" or reflected down into another place. TELSTAR enabled telephone and television signals to be transmitted internationally. Soon everyone had a phone. The next revolution came with mobile, or cellular, phones. With the widespread development of cell phones in the 1990s, people could take their conversations on the road. Today, there are approximately 6 billion cell phones in the world—a number not far behind the world's population of people (almost 7 billion).

ONE CALL THAT CHANGED IT ALL

With onlookers staring, Martin Cooper made a historic phone call from a New York City street in 1973. An engineer for the American telecommunications company Motorola, Cooper had just invented the first working cell phone. He called Joel Engel, the head of the cellular division at AT&T, who had also been trying to develop a cell phone. When Cooper told Engel he was calling on a mobile phone, the line went quiet. Cooper remembers: "My assumption was that he was grinding his teeth." Cooper's phone weighed more than 2 pounds (.9 kg) and cost $4,000 (equivalent to almost $20,000 today). In 1995, he won the Wharton Infosys Business Transformation Award for his contributions to communications technology.

Almost all new technological devices start out awkwardly large in size, including Martin Cooper's historic cell phone (opposite) and computers of the 1950s (left)

Computer technology, meanwhile, had taken off in the 1950s. The first computers were enormous machines that could cost hundreds of thousands of dollars to build. They could not do much more than a pocket calculator can today. But, initially, power was not the point. The significance of the first computer was that a machine—not a human—was doing the work. It was only a matter of time before parts got smaller, more efficient, and more affordable.

In 1969, the U.S. government started a computer network called ARPANET, which was the **forerunner** of the Internet. Although it was limited to a few users, the network let them exchange messages and information via computer. This was e-mail, the first—and still the most common—use of the Internet. By the 1990s, the Internet had gone public. It wasn't just for government agencies anymore, and companies and private individuals got into the act. David Clark, a computer scientist at the Massachusetts Institute of Technology (MIT), noted, "The great success of the Internet is not technical, but in human impact. Electronic mail may not be a wonderful advance in Computer Science, but it is a whole new way for people to communicate."

Although the terms are sometimes used interchangeably, the Internet and the World Wide Web are not the same thing. A British computer scientist, Tim Berners-Lee, was responsible for the "Web" on the Internet. In 1989, he began developing a system that allowed different pages of information to be linked so that users could click from one thing to another. The first decade of the Web is sometimes called Web 1.0. During that time, the Internet became a tool to get information. The second decade—the 2000s—was called Web 2.0. People could not only receive information but also create and post it. The digital revolution was well underway. It made existing forms of communication easier and more reliable. It also would end up changing the fundamental ways in which people communicated.

Communicative Headlines

MAKING THE LINK

In the 1980s, British software consultant Tim Berners-Lee wrote a computer program to help him remember the links between his colleagues and the projects on which they were working. Then he thought, what if information on computers everywhere could be linked and made available? Berners-Lee came up with software to do just that, and in 1991, he put his creation on the Internet, where it spread quickly. Today, Berners-Lee leads the World Wide Web Consortium (W3C), which develops Web technology and standards, as well as the World Wide Web Foundation, which works to make that technology available to more people.

Tim Berners-Lee (below) and other pioneers of the World Wide Web opened up a world of information to anyone with a computer, keyboard, and Internet connection

CHAPTER TWO 2 CHAPTER TWO

STAYING IN TOUCH

Today, consumers—the people who buy things—are demanding. We want phone calls that always go through, with the person on the other end sounding as if she's standing two feet (61 cm) away. We want our TV shows to be in high-definition so we feel as if we are standing right in the middle of the action. We want *more* information (quantity), and we want *better* information (quality).

Just as it was 150 years ago, when the Pony Express used the fastest horses to deliver mail, speed is still a benchmark for top-notch communication. Many mobile phone networks today use 3G (short for "third generation") or 4G technology. Users with advanced phones on certain networks can talk and send or receive data at the same time, and they can do it faster than ever before. Internet connections keep getting faster, too, partly to keep up with the demand for video. Video files contain huge amounts of data. Downloading a movie requires the same amount of **bandwidth** as sending 200,000 e-mails! In 2010, a new Internet **router** was introduced by Cisco Systems. It could handle 322 terabits per second—equal to downloading all the printed material in the U.S. Library of Congress in barely one second.

The availability of communications service has also improved. More satellites of higher quality have made it easier to bring Internet, television, and cell phone service to people in rural areas. However, satellites are not cheap, and most data is still carried by cables. In 2010, a new undersea cable was built between the U.S. and Japan. It could carry about five terabits of information—five trillion "bits" of digital information—per second.

Communications technology has advanced rapidly, with little more than a century separating the Pony Express era (left) and the age of communications satellites (above)

Increasingly, people can communicate with objects, controlling their homes' energy remotely (above) and receiving helpful feedback from sensor-equipped cars

While speed and availability allow for *more* information, variety addresses the demand for *better* information. Developments in **optical** character recognition and text-to-speech technology let people communicate in ways that once seemed impossible. For example, special computer software can "read" a book, then translate it into speech for blind people. Foreign-language translators let people enter a word or phrase into a **search engine** and have it returned in another language. Language barriers almost cease to exist with an Internet connection.

Computer scientists have also made breakthroughs in artificial intelligence, the process of making machines act intelligently—that is, as humans do. Although "thinking robots" are still far in the future, there have been developments that enable humans to communicate with objects—and vice versa. For example, an application for Apple's iPhone lets people stay in touch with their home. If they go on vacation and forget to turn off the lights, they can send an order from their phones to darken their house. High-end Mercedes-Benz cars come with **sensors** that collect various types of information. They can even detect whether their drivers are slumping (and perhaps falling asleep) and then wake them up. Some vehicles can also use sensors to locate objects around them and use the information to park themselves. Researchers in the Netherlands, Spain, and Turkey helped develop systems for cars that, in the event of an accident, would collect and transmit emergency information about passengers to hospitals. These systems could become standard equipment in luxury cars by 2018.

GPS technology lets people communicate with a whole network of information. Many people are no longer content to consult a paper map, let alone wander around lost while traveling. Many cars and cell phones now come with GPS technology that can give them directions to where they want to go. GPS is operated by the U.S. government but is available to the public. Russia has a similar system for its military, and China and the European Union have developed their own such systems. Even cowboys use GPS today. They put sensors on their cattle so they can track them remotely from a computer or cell phone. If the animals wander off, the cowboys can turn on an "electronic fence." Activation of the "fence" sends out electronically generated sounds that herd the cattle in the desired direction.

In some cases, the ability to pinpoint someone's location has been key to finding lost or kidnapped individuals (or cows). However, there is a flip side to this benefit. Finding someone through their cell phone, without their permission, could be an invasion of privacy. Some positioning systems not only identify a person's current location but also create a record of his movements, showing where he went and how much time he spent in each place. Parents can use GPS software to keep tabs on where their children are—a form of supervision that not all kids appreciate!

The advent of mass media allowed communication with large numbers of people, while the telephone and telegraph dissolved the obstacles of time and space. The Internet did both. The author, or

As a car navigational device (opposite), GPS can provide a convenient road map, and included in a phone (above), it can assist in determining a person's whereabouts

THE PARTY LINE

Keeping people connected can save time, money, and even lives. New devices and new ways of communicating aim to do all those things. People in remote or rural areas who used to be isolated can now receive medical treatment quickly in an emergency. Smart **power grids** *can help save energy. In a society that today values making connections all around the globe, communication is the first step. Of course, the corporations behind these efforts make a lot of money selling new technologies and service plans to use them, but for many people, the benefits—especially the sense of security—provided by always being in touch make it a price well worth paying.*

The Internet has become a buffet of information, entertainment, and communication options, offering easy access to the Web, social media sites, e-mail, and more

sender, could control when the message was sent, while the receiver could control when it was read. In addition, the Internet gave people access to more information. Taken together, these benefits have dramatically changed how people interact. To get the latest news right now, you need only click on a news Web site. If you're busy now, you can always read e-mail later. If you're feeling talkative, you can interact with friends or strangers via live chats. The Internet enables people to do all kinds of things online—shop, go to school, and even vote. With Web 2.0, the Internet let people be creators, too. They can comment on news stories or **blogs** and even design their own Web sites.

However, sometimes it's difficult to balance the benefits of in-person communication with the convenience of cyberspace. An e-mail exchange can get confusing when it includes too many people. So can a conference call. But scheduling everyone together can be difficult—and people don't always want to, anyway. "We want to communicate, but the methods we communicate with are just not in sync with our realities," says American Jeff Keni Pulver,

Printed newspapers (right), long a staple of information worldwide, are rapidly being replaced by online news sites (below), which can feature almost instant updates

a communications and Internet expert. Companies such as Sococo and Google have developed programs to address problems such as these. Using **avatars**, people can participate in real-time communication in **virtual** worlds that resemble their homes, offices, or a local coffee shop. For example, in 2010, Sococo introduced Team Space, which creates a virtual office environment.

By the end of the 20th century, mass media such as newspapers and TV stations recognized a shift in the way people preferred to get information. The Internet was giving people new notions about how they wanted to receive their news, when they would access it, and what they would learn. With traditional mass media, editors chose what their audience would read or hear. But the Internet was like a giant buffet of information, and people could select and ignore whatever they wanted.

TOO MUCH INFORMATION

With more and more digital information available, information overload is a concern. For many decades, people got news information only after it had been thoroughly reviewed, filtered, and packaged—through a newspaper or television broadcast, for example. However, these mass media outlets are losing ground to the Internet, and in the age of instant news, it has become harder to sort good information from bad. In addition, some experts worry that constantly communicating through devices such as computers or cell phones will make people devalue the advantages of old-fashioned, face-to-face interaction.

Millions of people—especially young people—communicate today via Twitter (right) and Facebook, the brainchild of programmer Mark Zuckerberg (opposite)

As the Internet has risen, mass media has declined. Fewer people today subscribe to printed newspapers or magazines. Television is not dead, but it's increasingly used more for sports and entertainment, while people turn to the Internet for news. However, some companies are working to merge television with the Internet, bringing together the video quality of television with the choice and interactivity of the Internet.

New technologies have even changed the basic ways people choose to communicate with each other. Teens and young adults often prefer texting on their cell phones instead of talking in person or sending e-mail. Many young people also spend a lot of their time on social networking sites such as Facebook or Twitter instead of hanging out in person. Even when they are together, they may still be on Facebook or texting each other. As of 2012, Facebook had nearly a billion users. Combined, these users spend the equivalent of 1.3 million years on the site every month! Mike Wesch, a professor at Kansas State University, explains this trend by noting, "This is as much a social or cultural revolution as it is a technological one."

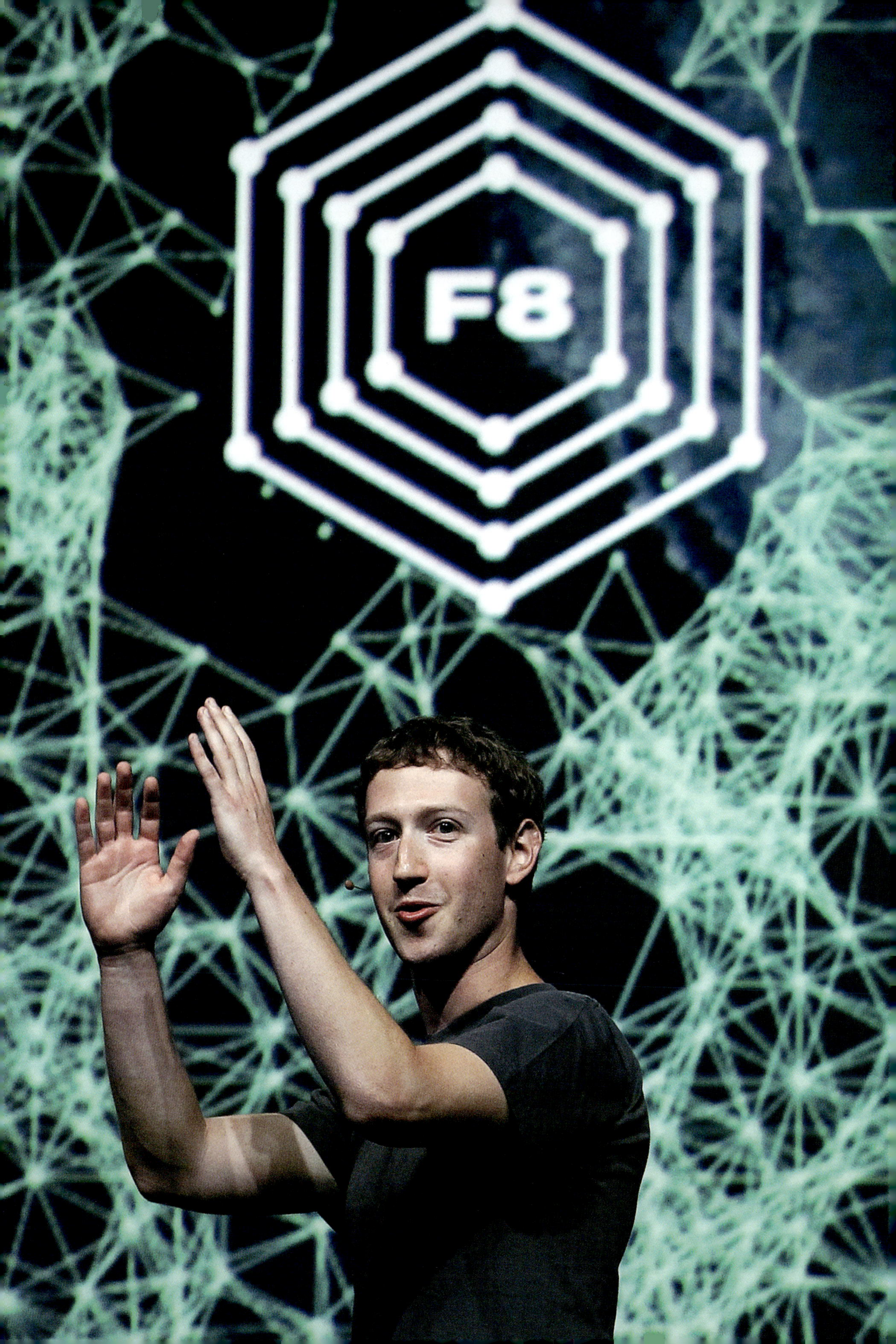
F8

CHAPTER THREE

3

SEMANTICS, SMART GRIDS, AND *STAR WARS*

D*o you know what I mean?* If half of communicating is exchanging a message, then the other half is understanding that message. However, sometimes things don't come through the way they were intended. People have always misinterpreted one another from time to time—and that's not likely to change, because everyone perceives things a little differently. However, machines can be programmed to "understand" things the same way every time. People may get frustrated talking to each other, but they could have better luck getting computers to see things their way.

Computer developers are currently thinking about Web 3.0, a time period that could usher in the "**Semantic** Web." Nova Spivack, an American technology entrepreneur, says it's time for the Web to take on another role. "We're about to enter the third decade—Web 3.0—which is about making the Web much smarter," he says. Currently, Web pages are designed for computers to find but for humans to read and understand. However, some computer scientists are working on software that would "teach" search engines how to interpret Web pages without humans having to get involved. A person would eventually read the page, but computers could save time by quickly sorting through mountains of information and returning only that which is most relevant.

The next step would be for devices to start acting like secretaries. Based on a user's history and preferences, a computer or cell

FOOD FOR THOUGHT

In American author M. T. Anderson's 2002 novel Feed, *the characters never have to think about what they want. Instead, a constant "feed" of information and advertising flows through their brains, suggesting what they should do, buy, and even think. The novel takes a dark look at the dangers of this kind of brainwashing, showing both its mental and physical effects. Advances in neuroscience are making it possible to manipulate the brain's natural processes and to interpret electrical brain waves. These developments are fueling an ethics debate about the extent to which the human brain can and should be manipulated in terms of communications technology.*

phone could take over some of the cyber-research. For example, a person might type "good restaurants" into a search engine. Well, what's good? In the future, a person's computer could "know," based on where the person has eaten in the past, that he likes Chinese food and doesn't want to spend too much money. It could return results suggesting places that met those criteria. On the other hand, someone else who typed in "good restaurants" might get a list of fancy Italian places.

As the Internet expands the ways it connects people to each other—as well as to places and ideas—the traditional roles of communicators may change. In a way, the Internet puts everyone on a level playing field. Anyone with an opinion and a connection can log on and make herself heard. Today, a person searching for "good restaurants" might get results based on what critics or other diners had to say. But in the future, this might be even more customized so that the person gets recommendations from her family and friends—the people she trusts to *really* know what she likes.

Logging on to the Internet is pretty easy. But it's still something a person has to make a deliberate choice to do. That might not always be the case. Researchers envision a future "Internet of things" that would connect everyday objects—and the people who use them—to a larger database. For example, clothing could come with sensors. These could measure a person's **physiological** data and alert him if something is wrong. If a person has a heart attack, it might be his shirt that automatically calls a doctor!

WHEN'S THE REVOLUTION?

The social networking site Facebook helped spur a revolution in Egypt in 2011. When an Egyptian citizen used his Facebook page to organize a public demonstration against the nation's president, thousands of protesters showed up. Gatherings like this are sometimes called "smart mobs." Unlike a traditional mob, communication allows people to plan their actions in advance. Modern communication technologies also make "flash mobs" possible. Flash mobs are organized in a very short time and often involve a silly, unimportant act. One of the first flash mobs was in New York City, when a magazine editor gathered people together at a store to pretend to be shopping for rugs.

With the world's population ever growing and its natural resources ever shrinking, clean and renewable forms of energy will become much more vital in the 21st century

Although people's clothes aren't yet wired, their homes are getting more connected than ever. The "Smart Grid," or "Intergrid," is a sort of Internet for power grids. These grids use computers to regulate how people use utilities such as electricity and gas. First, appliances such as refrigerators and clothes dryers are equipped with sensors that detect when people use them and how much power they consume. Then they send this information to the grid. Next, the system automatically orders the appliances to run at times when there is less demand. This makes it cheaper, because power often costs more during peak times. In addition, cleaner types of energy—such as wind—can also be used more often, cutting back on the use of traditional sources, such as coal, that cause more pollution. Already, smart meters are being installed in countries around the world, including the U.S., Canada, and nations in Europe. China is particularly enthusiastic, with plans to construct a nationwide grid by 2020.

Pants with pagers? Dishwashers with day planners? The idea of **integration** could play a big part in future communication. Just as people can now take pictures with their cell phones or listen to music on their computers, devices have the potential to become even more integrated so that nothing—and no one—is ever out of reach. R. David Lankes, a professor at Syracuse University, says, "What we're getting to is that point in the **Industrial Revolution** where people figured out that if you make the parts from machines interchangeable, you change the world."

We remain a long way from the technological capabilities featured in shows like *Star Trek* (right), but the speeds at which we can share data continue to improve

It has been said that the human brain is the ultimate computer. Integration could truly come to a "head" when man-made computers are linked to biological ones. Computer chips implanted in people's brains might be able to improve their vision or memory. A brain-centered GPS would make sure someone never got lost again. Brain chips might even change how people think. According to American inventor Ray Kurzweil, adding non-biological intelligence to people's brains will become a routine upgrade. "In 2035, you're not going to be able to walk into a room and say, 'humans on the right side, machines on the left,'" he says. "It's going to be all mixed up and integrated—one complex, dynamic, chaotic human/machine civilization." Other scientists disagree with Kurzweil, arguing that what he envisions is basically a robot, and that there are too many technical challenges and ethical issues standing in the way for his predictions to come true—at least not any time soon.

Integration and connectivity are **buzzwords** in communication, but other advances will aim to tackle the traditional barriers of space and time. In the popular television show *Star Trek*, Captain Kirk (and a few lucky crew members) were able to "beam" themselves through outer space in a few moments. There are some major physics problems with **teleportation**, however, so researchers are working on the next-best solution: holograms. This was one of the ways characters in another popular science-fiction saga, *Star Wars*, stayed in touch.

Holograms use light to reconstruct recorded pictures and create three-dimensional images. The result is an image that suggests a person is actually present. Today's holograms can be grainy-looking, but developments in light technology, as well as advances in the computer speeds needed to reproduce 3-D images, will make them more realistic.

The creation of realistic holograms, the redesign of the World Wide Web to make information easier to find, and the development of faster phone connections all share a basic need: computing power. Machines in the future—whether they are supercomputers, cars, or lowly toasters—will be able to do more than ever before. They will be able to do it faster and more cheaply. Even more importantly, they may be able to do it differently. Computers now depend on small electrical components such as transistors and semiconductors to transmit signals, but future technology will likely move beyond those. For example, U.S. researchers at IBM have found a way to convert light into electrical signals. This could lead to the transference of more data—and at much faster speeds—than can be done using wires. Researchers in Britain have harnessed the human body's electrical and chemical systems, using **DNA** to create electronic switches that might be used in future devices. Technology has the potential to set new standards for how people send, receive, find, and process information. The next step will be for people to decide which new devices and forms of communication will improve their lives—and which won't.

CHAPTER FOUR 4

LEARNING THE LANGUAGE

With the development of cell phones and instant messaging, a new "language" was invented: text speak. This language uses combinations of letters and numbers to represent entire words, such as "u" for "you," or "l8r" for "later." It evolved because it was well-suited to technology that depended on writing—which is slow—but offered the immediacy of talking—which is fast.

Yet another technological language could evolve with Web 3.0. For an idea such as the Semantic Web to work, computers will have to recognize and translate data on Web pages—a job that is now left to humans. Software will be needed to create and interpret "metadata," which is data that describes other data. However, this might mean using a whole other computer language, one that is more difficult than the HTML that is used for most Web sites today.

Critics say that making this transition across the entire Internet would require tremendous amounts of work. The change also would come with problems, such as teaching computers what to do when contradictions or inconsistencies come up. If they are looking at billions of pages originally created by humans, there are bound to be a few! In addition, the market for the Semantic Web is unclear. What types of tasks could really be done much better and faster—and would people trust the computer to do a better job than they could do themselves?

It costs money to buy a computer and obtain Internet service, but once you've done so, most of the Internet's information is free

Understandable language is at the heart of any form of communication; the Web "understands" HTML (bottom), while texters have their own language (right)

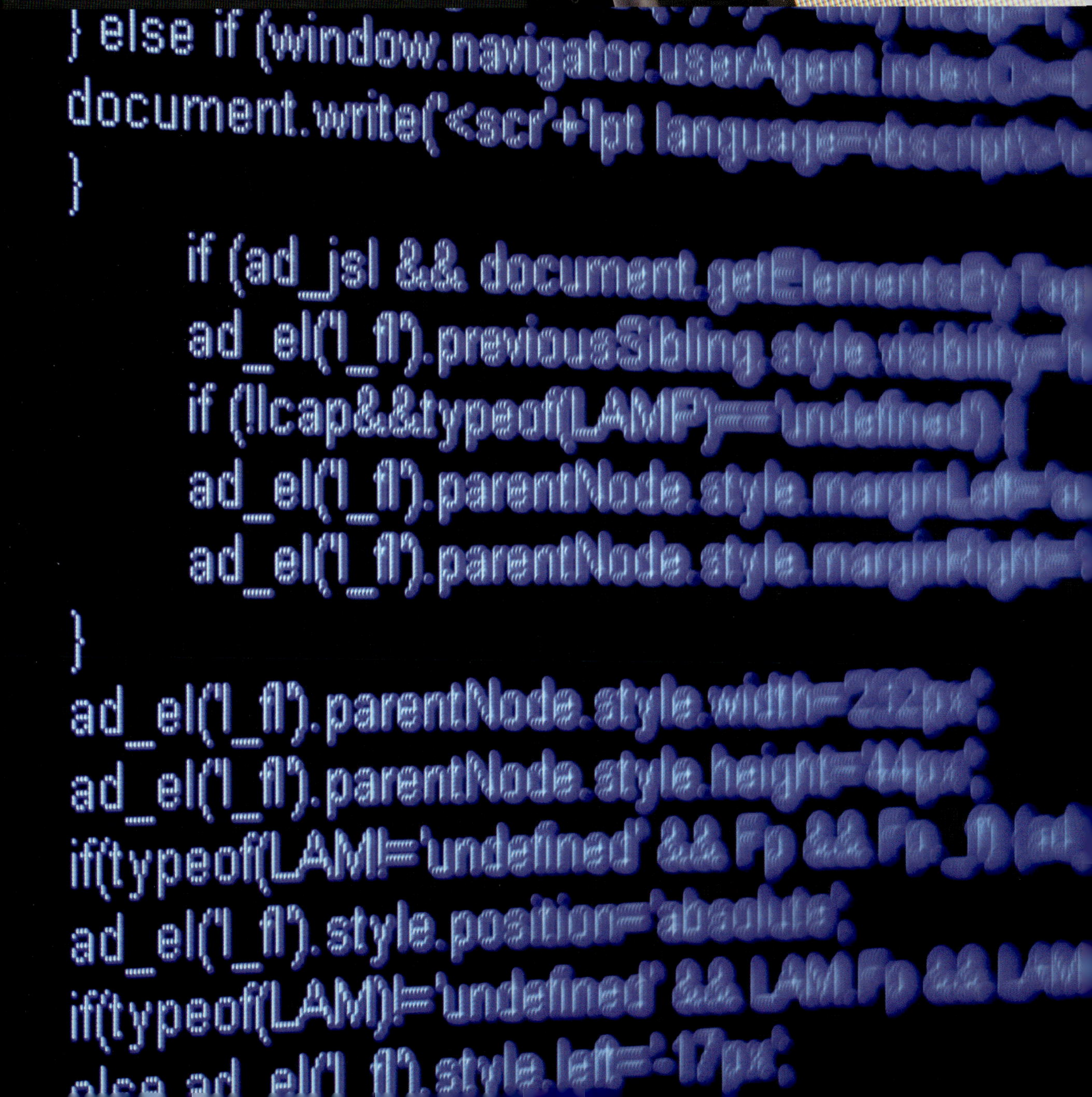

ON EQUAL FOOTING

A hot topic in Internet management today is "Net neutrality." Currently, the Internet is structured so that all Web content is equal in terms of accessibility. It's just as easy to get international news from a big company such as Google or Yahoo as it is to shop at a tiny online store selling handmade oven mitts. However, some large communications companies argue that if they are the ones providing access to the Internet, then they should be able to promote their content by making it more easily available. However, in 2010, America's Federal Communications Commission (FCC) approved new regulations that would make it more difficult for companies to limit access to competing Web sites.

and easily accessible. However, that may change if companies decide to charge money to access content on the Web or make it so that it's available only by using certain devices or software that a consumer would have to buy first. Already, some content is fee-based or can be viewed only by members who sign up. Territory is getting divided up, prompting one journalist to call it the "Splinternet." Politics are another issue. For example, China's government has the "Golden Shield Project," which blocks Chinese computers from accessing certain Internet sites. (The government does this largely to prevent citizens from seeing alternate political views.) Australia plans to set up a similar system to block pornography.

Privacy is another Internet issue that is likely to generate discussion in the future. As more and more people go online, computer networks are collecting all kinds of information about them, from what kind of shoes they buy to how much money they earn. Built-in software programs called **cookies** collect this information when people enter it on a Web site. Computer programs called **spyware** can also be secretly installed on someone's computer over the Internet and collect information they did not mean to reveal. Even when a person provides information on purpose, however, it's hard to take back, and it can be used in ways a person didn't intend.

Meanwhile, traditional concerns regarding **plagiarism** and copyright infringement need to be considered. Digital content is extremely easy to access, copy, and distribute. Already, some people question whether laws about copyright—protecting intellectual property—should be changed to reflect these new realities. Accuracy is an issue because, unlike traditional media such as newspapers, the Internet has few filters. Anyone can go online and write about anything—whether it's true, false, or somewhere in between. Unfortunately, readers often won't know which is which.

While Facebook (right) spreads its social networking influence worldwide, some developing countries still run pay-per-use telephone access service stations

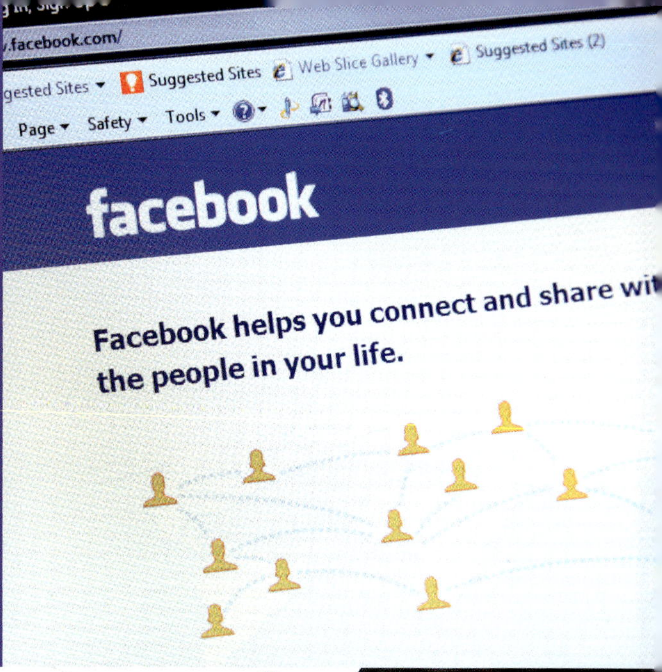

THE MAGIC NUMBER

The magic number is about 150, according to British researcher Robin Dunbar. Based on the size of the human brain, Dunbar theorized that this was the maximum number of people with whom any one person could have meaningful social relationships. He believed the human ability to speak made this number larger than that of animals, which must depend on time-consuming physical contact such as playing or grooming. With social networking sites, however, some people think that this number could climb higher, since it takes less time to connect with people virtually. Interestingly, according to Facebook, the average user has links to about 130 friends—a number quite close to Dunbar's.

Finally, there is the problem of access. In **developed nations**, millions of people use the latest communications technology, but in poorer countries, millions more do not. The "digital divide" refers to the gap between people who have access to advanced technology and those who do not. Even in developed nations, younger people who have grown up with technology are far more comfortable with it than older people who sometimes struggle to master it. As new communications technologies become more common, it could leave some people behind. American communications expert Terry McKenzie says, "The whole notion of digital natives and digital immigrants [is] very relevant at this point. For people who grew up in the digital age, it's not going to be a big deal. For digital immigrants, it's going to be much harder."

Whether it's the Internet or cell phones, there are technical issues, as well, that will need to be resolved to allow for future improvements. With billions of pieces of information flying through cyberspace every second, there are two main challenges in keeping traffic from getting snarled. There need to be more and better "roads"—such as cables, satellites, and routers—to handle all the data. And data needs to move along these roads at ever-faster speeds.

Some experts believe that the future of communication will be centered around cell phones, since they have the potential to pair computing power with mobility. What will be necessary to live life from a handset? For this to work, the communications channels for

Even many giants in the field of computer technology, including Gordon Moore (opposite), have been startled by the world's rapid arrival in the 4G era

cell phone signals are going to have to increase dramatically. Even as some cell phone companies are still upgrading to 3G technology, 4G (fourth generation) is already here. And even that may not be enough. Multipurpose phones such as Apple's iPhone are leading users to perform more data-intensive actions, such as downloading video. According to Roger Entner, a former senior vice president at the Nielsen consumer research firm, "4G is not a **panacea**. You could very well see the same problems you've seen on 3G with 4G." Even though 4G networks move faster and can carry more information, they could get maxed out as people's demands continue to increase. "It's just a matter of how long it takes to fill the 4G network," Entner adds.

Fortunately, technical advancements in computing are continuing at astonishing rates. Back in 1965, Gordon Moore, cofounder of the computer company Intel, noticed that computer hardware was quickly shrinking in size while growing in capacity. He predicted that trend would last another 10 years. He was both right and wrong. It did last another 10 years. In fact, it's lasted almost 50, and it's still going.

Although traditional computing is at the heart of the digital revolution, future developments will require advances in other branches of science and technology. For example, nanotechnology—which involves the creation and manipulation of tiny particles of matter—will probably spur the development of the "Internet of things." Light technology will be critical for developing holograms and improving the visual quality of screens on televisions and computers. Meanwhile,

in artificial intelligence (AI), computing will meet biology as scientists work to create "**neural** networks" that can imitate the way the human brain functions and help machines do the same. Already, doctors have implanted tiny computer chips into people's brains to help with medical conditions such as Parkinson's disease. As this technology advances, scientists will need to determine whether computers would communicate with the brain or actually upload certain skills or memories. This brings up ethical questions, such as whether rich people could pay to be made "smarter."

This was not a question people would even have dreamed of asking 50 years ago. Improved technology has given people more possibilities than ever before and more choices to make. Most people did not want a telephone before they'd ever heard of them. They did not realize they could not live without e-mail until they learned what it could do. People may not always know the right things to say to each other, but whatever they decide, it's certain that in the future, they'll have plenty of ways to say them.

GLOSSARY

avatars — pictures or symbols used for identification in digital environments

bandwidth — an amount of space used for transferring digital data

blogs — logs or journals created and distributed online; the term is short for "weblogs"

buzzwords — popular words or expressions intended to evoke certain reactions

cookies — types of software used to automatically collect information that people enter on Web sites

developed nations — the wealthiest countries of the world, which are generally characterized by good health care, nutrition, education, and industry

DNA — an organism's unique combination of genes, which are hereditary units that determine the organism's particular characteristics

forerunner — something that came before something else, often contributing to the latter's achievement

Industrial Revolution — a period from the late 1700s through the 1800s when machinery became commonplace in developed countries such as Britain and the U.S.

integration — the process of combining two or more things in order to maximize their abilities or efficiency

intrusive — invasive, or present in an unwelcome way

medieval — relating to the Middle Ages, which spanned the years from approximately A.D. 476 to 1453; a time before scientific knowledge and technology became widespread

mogul — an important and powerful person, especially in a particular industry such as media

neural — relating to the nervous system, especially the brain

optical — relating to eyes or the sense of sight

panacea — a cure-all, or a treatment used broadly to solve many problems

physiological — relating to the body's physical functions and processes

plagiarism — the act of copying material created by someone else without giving credit for it

power grids — networks of equipment that supply electrical power to people

router — a device that controls the incoming and outgoing flow of information to a computer network

satellite — an object placed into Earth's orbit to receive and transmit information over long distances

search engine — a computer program used to search for information on the Internet

semantic — relating to the meaning of words or language

sensors — objects that can pick up and record data

spyware — software that is installed on a computer without the owner's knowledge and collects information about his or her Internet activities

teleportation — the process (currently impossible) of moving an object instantaneously through space

venture — an activity or project, often containing some degree of risk

virtual — relating to something done outside the physical realm but meant to achieve the effects of something done in reality

SELECTED BIBLIOGRAPHY

Baumann, Michael. "Web 3.0: The Next Step for the Internet." *Information Today*, May 2009.

Cohen, Aaron M. "Merging with the Machines: Information Technology, Artificial Intelligence, and the Law of Exponential Growth." *World Future Review*, April/May 2010.

Crow, Barbara, Michael Longford, and Kim Sawchuk, eds. *The Wireless Spectrum: The Politics, Practices, and Poetics of Mobile Media*. Toronto: University of Toronto Press, 2010.

Schiller, Kurt. "Five Trends Shaking Up the Internet." *Information Today*, October 2010.

"A Virtual Counter-Revolution." *The Economist*, September 4, 2010.

Webb, William. *Wireless Communications: The Future*. Hoboken, N.J.: John Wiley, 2007.

Willsky, Kate. "Plotting a Course for the Future: What Can Communicators Expect to See over the Next Decade?" *Communication World*, May 1, 2010.

WEB SITES

Answers for Young People

http://www.w3.org/People/Berners-Lee/Kids.html

On this page, World Wide Web founder Tim Berners-Lee provides information about himself and the Web in a question-and-answer format aimed at students.

HowStuffWorks: How Holograms Work

http://science.howstuffworks.com/hologram.htm

This site provides a detailed but accessible explanation of the science behind holograms, exploring how light waves, lasers, lenses, and mirrors can all play a part in the process.

INDEX

Published by Creative Education
P.O. Box 227, Mankato, Minnesota 56002
Creative Education is an imprint of The Creative Company
www.thecreativecompany.us

Design and production by The Design Lab
Art direction by Rita Marshall
Printed in the United States of America

Photographs by Alamy (Everett Collection Inc., Janine Wiedel Photolibrary, Photos 12, Chris Ridley-Internet Stock), Corbis (Eloy Alonso/Reuters, Andrew Brusso, Rick Friedman), Dreamstime (Alexstar, Beisea, Sebastian Czapnik, Djembe, Melinda Fawver, Frank Gärtner, Lucian Milasan, Mozzyb, Orangeornate, Photowitch, Picstudio, Oleg Pidodnya, Rgilman, Scattoselvaggio, Roman Smitko, Johannes Gerhardus Swanepoel, Dennis Thompson, Alexey Utemov, Hannu Viitanen), Getty Images (Buena Vista Images, commerceandculturestock, Bert Hardy, Kazuhiro Nogi/AFP, Justin Sullivan, Three Lions, Time Life Pictures/Mansell), iStockphoto (alengo, Hulton Archive, Merve Karahan), Library of Congress, Shutterstock (IgorGolovniov, italianestro, Stepan Kapl)

Library of Congress Cataloging-in-Publication Data

Bailey, Diane, 1966-
The future of communication / by Diane Bailey.
p. cm. — (What's next?)
Includes bibliographical references and index.
Summary: A look at potential future developments in communication, including holograms, as well as the Global Positioning System and other technologies that are currently considered state-of-the-art.
ISBN 978-1-60818-221-3
1. Telecommunication—Forecasting. 2. Telecommunication—Technological innovations. I. Title.

TK5102.4.B336 2012
384—dc23 2011040506

First edition

9 8 7 6 5 4 3 2 1

Cover: A satellite view of Earth
Page 1: An Ethiopian boy using a laptop computer
Page 2: Satellite antennae on a telecommunications tower